the **up**cycled garden

the **up**cycled garden

25 step-by-step projects using reclaimed materials

Steven Wooster

Susan Berry

A Berry publication

First published in 2016 by
Berry & Co (Publishing) Ltd
47 Crewys Road
Childs Hill
London NW2 2AU
www.berrypublishing.co.uk

Designer **Steven Wooster**
Editors **Susan Berry**, **Katie
Hardwicke**

British Library Cataloguing in
Publication Data
A catalogue record of this book is
available from the British Library.
ISBN 978-0-9927968-2-2.

Reproduced in UK
Printed in China

MIX
Paper from
responsible sources
FSC® C010256
FSC
www.fsc.org

SCREENS & SUPPORTS

CONTAINERS & RAISED BEDS

INTRODUCTION

We had both acquired allotments on the same site in London and became fascinated by the diverse and curious constructions that plot holders had created from materials that were to hand. We also realized we needed some similar structures on our own plots. Between us, we also have three gardens: two in London, and one in France, all in need of improvement and all with room for some additional features, so there was plenty of space, and scope for different kinds of project for this book. So where did the ideas for the different structures come from? Steven became interested after doing a garden design course at Capel Manor College a few years ago and then built simplified garden structures in two separate garden designs that he did for the Conceptual Gardens category at the Hampton Court Flower Show in London.

But the main reason we both like using recycled materials is that, firstly, doing so makes good use of items that would otherwise go into landfill and, secondly, the materials themselves present interesting practical and design challenges. It is important to keep the essence of the original materials in mind when reusing them: in other words, don't try to make the project look too slick but, conversely, don't just attach four legs to a pallet and call it a coffee table!

There is an emphasis on wood-based projects in this book as wood is generally more readily available and is easier to work with, especially for the novice builder, although we have included other materials, as and when we found suitable ones.

Opposite A simply built but effective potting shed made from an assortment of timber planks and offcuts. The window was a salvaged piece as well.

All the projects in this book were made by Steven on our respective allotments or gardens. In some cases, the projects simply recycled a structure that was already there, but in a poor state of repair (for instance Susan's easy-build shed and the revamped wooden deck). Others were in response to a need: for example, Steven needed a greenhouse in his London garden and the lure of the discarded leaded windows was too great to resist.

You don't need to be purist about using only recycled materials. You may occasionally make it easier for yourself if you buy a few items new, such as studwork, joists, strand board (osb) timber, roofing felt and so on.

As Steven constructed the projects, he took step-by-step photographs and made notes as to the methods used. Because it is not possible for anyone to copy these projects exactly (apart from the Adirondack chair), we have avoided exact measurements of the elements involved in the structures, as you will be using your own found materials to create similar projects.

Most of the projects are easily made by one person but, depending on your strength and the size of the project, you may need a friend to act as building mate when lifting heavy items or holding more cumbersome pieces in place.

Above A gently meandering brick path made from recycled bricks was the inspiration for the reclaimed brick circle (see page 40).

The projects in this book are graded in terms of skill level and/or time taken:

SKILL LEVEL

SKILL LEVEL

SKILL LEVEL

1 No previous skills needed

2 Some basic skills needed; simple DIY level

3 Advanced skills needed; experienced DIY level

- -

Tips on sourcing materials

■ Always keep a lookout for materials: skips are a good source, but always ask permission before removing anything. Remember that you will need a place to store materials so that you can stockpile the items over time.

■ Pallets come in various sizes, strengths and weights. Some come apart quite easily (see page 10), others can be more or less impossible to take apart. Unfortunately you cannot check first! If you have pallets that you can't prise apart easily, you could use them for a planted pallet wall instead (see page 92).

■ Try to develop an eye for industrial materials and items that may fit another purpose – for example, plastic guttering could be cut into lengths to make a shallow planter (page 116) and plastic tubing could be cut into short sections for a bug hotel (page 136).

Basic skills

Here are a few basic skills that will make the projects in this book quicker and easier to construct. Most of the tools used are readily available and may be part of your tool kit already. (See also page 138 for information on tools.)

Working with pallets

Pallets are a good source of free timber so knowing how to work with them and how to take them apart can be a big advantage.

1 Working from the middle of the pallet, begin by prising up a slat with a crowbar..

3 To remove the slat, apply more leverage by inserting the end of a long, stout timber batten under it.

2 When a slat has been lifted slightly, reposition the crowbar close to the nails, to remove or loosen them.

4 Continue by lifting each end of the wooden slats. But if the slats refuse to budge, your other option is to saw off each end (see opposite, top).

Cutting slats with a saw

If nails cannot be removed easily (often a problem at pallet ends), rather than forcing out the nails and possibly splitting the wood, saw the slats off at the ends instead (below).

Removing nails

Use a claw hammer to remove nails (below). If pallets are held together with nails with a annular (ringed) shank that have rusted, you may not be able to lever them up. Your other option is to hammer them in, but be really careful not to saw through a hidden nail, especially with a power saw!

Cutting timber

For cutting rough timber (such as pallet wood) across the grain use a saw with coarser teeth. (Saws are graded according to teeth per cm of blade.) To achieve an accurate cutting line, many saws have shaped handles that create a 45 or 90 degree angle (below) when laid on the timber to be cut. If you don't have a marker pen or pencil to hand, score a line on the timber with a nail point. If your saw sticks in the timber when cutting, run a bar of soap or a candle along either side of the saw blade.

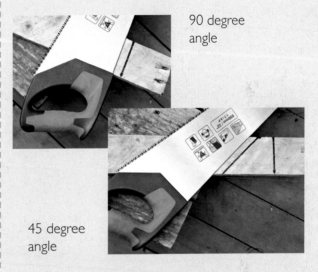

90 degree angle

45 degree angle

Cutting timber along the grain

Hand saws cut best across the wood grain. To cut along the grain, use an electric jigsaw (below), carefully following a previously marked line drawn on the wood. Use a coarse-toothed blade and make sure the timber is held firmly in place.

Cutting curves in timber

Some projects in this book require circles or curved sections of timber. A plate or a pair of home-made compasses can be used to achieve the required size.

Draw around a plate to mark out a circle.

I To make a simple pair of compasses you need a thin batten, a nail and a pencil, which you insert through a pre-drilled hole in the batten.

2 Hammer the nail lightly into the timber to be cut and, holding the other end, inscribe a circle using the pencil.

3 Follow the marked line to cut the circle out of the timber using a jigsaw with a narrow blade.

Cutting small holes in timber

For the concrete candle holder project (see page 28) you need to cut small circles of wood.

I Use a drill attachment like the one shown above.

2 Hold the timber firmly and drill halfway through it.

3 Then turn the timber over (above) and drill again. Insert the drill bit in the central hole. As you drill, the drill bit gets hot so take care when removing the timber circle.

Cutting metal

A metal sheet was cut from a tin can for the rocket stove project (see page 36).

1 Cut an empty can or biscuit tin using snips. Wear a sturdy glove as the cut metal will have sharp edges.

2 Fold over the edges to strengthen the metal and make it safer to handle.

3 The finished tin sheet is ready to be drilled.

Cutting a brick in half

A half-brick is used in the rocket stove and they are also used in the reclaimed brick circle (see page 40). Modern pavier bricks will cut more easily and neatly than old house bricks.

1 Using another brick as a measure, scribe the brick to be cut with a cold chisel at the halfway point on all four faces.

2 With moderate force apply a bolster hammer to the cold chisel. Do this on all four faces of the brick.

3 After a few blows on all four faces the brick will split cleanly in half.

FURNITURE & ACCESSORIES

Planted table

Adirondack chair

Revamping a stool

Tealight holders

Pallet occasional table

Rocket stove

Planted table

As soon as I found the pallet below
I knew it had the perfect configuration
for making a planted table - a
popular item at recent garden shows.
All that was required was to cut it
down to size and to create planting
pockets. As the wooden slats were
of lightweight construction, they
would be easy to remove: heavier-duty
pallets can be much more difficult,
if not impossible, to take apart.

SIZE 1000mm (w) x 540mm (d) x 300mm (h)

Right Sedums, being shallow-rooted, make
the ideal choice for this kind of planting – they
don't require constant watering or much
attention. Crushed roof tiles have been used
here as a decorative mulch.

Constructing the table

Start by cutting your table to your chosen size, then add the legs and the planting pockets, as required. This table has three pockets along one side.

Materials
Pallet, battens, butyl liner or thick plastic sheeting, screws, nails, staples, tacks or drawing pins.

Making the table top

1 Using a coarse-toothed saw, cut your pallet roughly to size.

Making planting compartments

4 These can be made using cross-pieces from the pallet.

2 Use slats from the off-cut piece of pallet to fill in the table top. Fix slats with screws or annular nails.

3 Cut more slats as needed to fit, making a level table top, leaving spaces for planting pockets.

5 Cut timber to fit. This shows a cross-piece to make a planting pocket.

6 Add slats to the underside of the table. These will support the planting pocket liner when filled with soil.

7 The finished planting pocket, with timber surround on all four sides.

8 Add legs cut to a height to suit. Screw each leg into the cross members at each corner for greater stability.

Finishing off

9 Line the pockets with thick plastic or an offcut of butyl liner. It can be held in position with drawing pins or tacks, or even staples. Puncture the base of the liner to allow water to drain away freely.

10 For a neat finish, add a piece of timber to each end of the table to hide the ends of the cross-pieces.

Left The table is ideal for seasonal plantings. In summer, it houses a cool oasis of nasturtiums, lobelia and dwarf dahlias. Another good planting combination would be a mixture of herbs – with a sprig always at hand to add to your favourite drink!

Adirondack chairs

2

SKILL
LEVEL

I chose to make an Adirondack chair primarily for the ease of construction, but also because it looks good – elegant and not too clumpy – and it can be made from flat boards. I had a supply of them left over from a recent decking project. With a little practice, you can make chairs with rounded seats or with contoured backs. The chairs can be stained or painted (see opposite) in a colour of your choice.

SIZE 650mm (w) x 900mm (d) x 1100mm (h)

Above and **opposite** The chair on the right (above) has been left untreated and over time will soften to a silvery-grey; the others are all painted. Altering the length of the angled legs will change the seat height and the rake of the back. (The Adirondack chair is named after the mountains in upstate New York, where it originated.)

Constructing the chair

Start by cutting the five main pieces to create the front cross-member, the two upright front legs and the two angled legs.

Materials

Templates (see pages 138-9), timber boards, nails and screws. Paint or wood dye as required.

Making the chair base

1 Begin by attaching the two upright legs to the cross member. The cross-member should be the width of five boards plus the gap of 6mm between each pair. This allows the back to fit neatly without cutting any boards lengthways.

3 Attach the boards for the seat, leaving an even 6mm gap between each board.

2 Screw the two angled legs into position on the two upright front legs.

4 Add a batten to the underside of the seat, to which the seat back will be screwed later.

Making the chair back

5 Cut five boards to the chosen length for the back plus a little extra for trimming to size later. Nail two cross-pieces into place to hold the boards together, but don't drive in the nail heads fully as this is only temporary.

6 Screw the back section into the batten fixed to the underside of the seat. The rebated back panel should fit neatly between the leg pieces.

7 Cut a cross-piece slightly shorter than the width of the chair back and screw it into place. Then remove the two (lower) temporary boards by gently levering out the nails.

Adding the arms and finishing off

8 Using templates, cut two boards to make the arm pieces that hold the chair together. The curved shapes are best cut with a jigsaw. The two smaller pieces act as brackets for the arms.

9 Fix the brackets to the leg uprights and screw the arms into the brackets and the chair back. Add two seat boards behind the back to strengthen the structure.

Revamped stool

Two old stools with chrome legs were updated by polishing up the legs and replacing the old seats with new ones from plywood and foam (the latter cut to size by the supplier), covered with sacking or similar.

Materials
Plywood, foam rubber, plastic sheet, sacking, staples.

SIZE **300mm (w) x 300mm (d) x 500mm (h)**

Remaking the stools

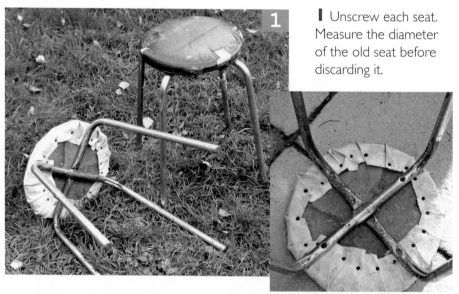

1 Unscrew each seat. Measure the diameter of the old seat before discarding it.

2 Remove any flaking chrome from the legs with a scouring pad dipped in oil, such as WD40.

3 Polish up the chrome using a soft rag.

4 Using a jigsaw, cut two plywood circles to the required size (see page 12).

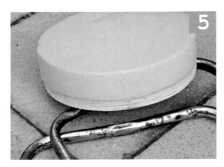

5 Place foam rubber pads cut to the same size as the plywood circles on top of each circle.

6 Cut a circle from the chosen cover and the plastic sheet about 15cm (6in) bigger all round. Invert the foam and ply circle on top. Staple the new cover around the ply base..

Above The re-assembled stools, covered with sacking, and with the polished-up legs screwed to the base of each new seat. As an alternative to sacking, oiled tablecloth fabric would makes a good waterproof covering for the seats.

Tealight holders

SKILL LEVEL 1

Tealights are a great addition to any garden or table setting, but they need some kind of casing as they become very hot and can be easily overturned. You could put them in small jars but individual concrete holders are easy to make and don't get hot. The ones shown were made using a wooden mould to shape the concrete, although you could use other forms of mould (see overleaf.

SIZE **90mm (w) x 90mm (d) x 50mm (h)**

Above Tealights positioned each side of an uneven path help to show the way. Never leave tealights unattended and don't use naked flames in areas of high fire risk.

Right The same holders are used to decorate a table at dusk. Citronella-scented tealights have the benefit of deterring mosquitoes.

Making the moulds

1 Cut out circles of wood from a piece of timber – when cut, the circles should be of a similar depth and just a fraction wider than each proposed tealight. Cut the circles with a large drill attachment (as shown). The wooden circles can get hot so take care when removing them from the drill.

2 Decide on the size/number of your tealight holders and draw out a grid on a thin sheet of timber. Allow for the width of the spacer battens and the frame. Mark the centre of each square. (The frame shown has nine holders but you can adjust to suit.)

3 Drive a short nail (roofing or clout nails are ideal) though the centre of each square and place the cut wooden circles on them as shown to prevent them moving when the frame is filled with concrete.

4 Cut four battens to the size of the grid and screw them together. The frame should also be screwed to the base. Using screws will make the frame easier to take apart later.

5 Cut dividers as shown and screw the end of each section to the frame. The frame should be made of battens of the same depth.

6 Carefully fill the frame with a concrete mix, tamp down the mixture and level off the surface. Allow to set for at least 12 hours.

7 When set, invert the frame and remove the base. If the wooden circles are set very firmly in the concrete, place two screws in each one and use pliers to twist the circles out. Then unscrew the frame and carefully release the blocks. The frame can be reused time after time.

Left If you just want to make a few holders, then old plastic food containers can be used as a mould. Screw in the wooden circles to keep them in position.

Pallet coffee table

2

SKILL
LEVEL

A pallet is naturally a perfect shape for a coffee table to use in the garden or in the house – think loft living. Wooden or recycled metal legs (used in this example) can be easily attached, and you can add large castors to make the table more manoeuvrable. The space between the pallet sides will accommodate easily made drawers, perfect for storing plates and cutlery.

SIZE 1100mm (w) x 440mm (d) x 320mm (h)

Materials
Pallet, table legs, scrap plywood, screws, nails.

Right The table was made from the pallet (above). Simple handles for the drawers were made from cut pallet pieces, but you could use any type of recycled handle if you prefer.

Constructing the table

Making the table top

1 Decide how big you want your table to be and carefully remove any surplus planks from your pallet (see page 10).

2 Cut the pallet to the desired size. A large pallet table would be heavy, in which case it might be best mounted on castors to make it easier to move.

3 Saw any removed planks to the required width – this is most easily done with a jigsaw or a circular saw. Then position as shown to construct your table top. Nailing the new planks will give a more coherent look.

Adding the legs

4 Mark the positions for the legs. These had bolts so a recess was needed .

5 Position legs as required and mark out section to be drilled using a large drill bit.

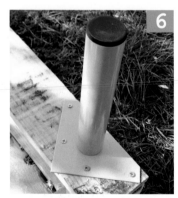

6 Screw legs into place. If painting, do this before fixing.

Making the drawers

7 Measure the area of the pallet for the drawers and then make them to fit using pallet planks attached to a thin plywood base. Screw the sides together first and then screw them to the base. Note: Always measure the pallet pieces for each drawer carefully as sections of pallet aren't always of equal dimensions!

8 Check the drawers fit snugly and plane them smooth if necessary.

9 Add planks, sawn to size, to the open ends of the table and fix in place.

10 As a finishing touch, to convert a drawer for cutlery, simply add some infills made of pallet wood. Screw the infill sections into place. Fix a thin piece of board to the front of each drawer (large enough to cover any gaps) and attach a handle to each drawer front.

Rocket stove

If you want to cook a simple meal or boil some water for a hot drink, a rocket stove can be made very simply from 15 old paving bricks and a sheet of perforated tin (see page 13).

SIZE **300mm (w) x 300mm (d) x 300mm (h)**

Materials
Paving bricks, broken tiles, piece of tin, tin snips.

Left Bacon and eggs sizzling away for breakfast on the rocket stove. You can use a few broken tiles to alter the height of the pan from the heat source. For this project just one brick (above) needs to be cut in half (see instructions on page 13).

Building the stove

1 Carefully position three whole bricks and one half brick (see page 13) to form a square, with a central gap.

2 Use an old metal biscuit tin to make a rack for the kindling.

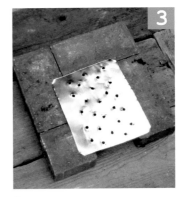

3 Using tin snips, cut a rectangle of tin large enough to cover the gap between the bricks, and perforate it.

4 Place a second layer (three whole bricks and one half-brick) on the first layer.

5 Create a third layer, this time using four whole bricks. Stagger the joints.

6 Add a fourth layer of bricks to complete the rocket stove, continuing to stagger the joints as you would if making a brick wall.

7 To light the stove, insert kindling through the gap at the base (to sit on the metal rack). Light with a long taper and use a firelighter if needs be. Add a few broken tiles to adjust the height of the pan from the heat if required.

STRUCTURES & SURFACES

Reclaimed brick circle

Easy-build shed

Reviving a deck

Open pallet gazebo

Log path

Retro greenhouse

Playhouse den

Reclaimed brick circle

2

**SKILL
LEVEL**

A brick circle set into a lawn makes an ideal space for growing herbs and small flowering plants. Here, reclaimed bricks are used to give a cottage-garden look that is more forgiving if your geometry goes a bit awry. If bricks are collected from various sources you need to mix the hues and colours, rather than place blocks of similarly coloured bricks together.

SIZE **1650mm diameter circle**

Above and **opposite**
The brick circle planted up for late summer. Plants include herbs such as French lavender, geraniums, thyme and mint, dwarf dahlias and platycodons.

Making the brick circle

You need to mark out the size of the circle and the central planting hole before excavating the soil to the required depth, and adding a sharp sand base. Lay the bricks from the centre outwards. Then plant up.

Materials
Reclaimed bricks, sharp sand and ballast, sticks and string, wood batten.

Marking out

1 Position bricks to determine the radius of the circle. Use a stick (or parasol base) for the centre point. Tie string to it with a sharp point at the other end, and inscribe a circle slightly wider than the brick paved area.

2 To prevent your circle from being obliterated when you dig out the soil for the base, go over the circumference, marking it with pale water-based paint.

Digging out

3 Dig to a depth just deeper than brick height.

4 The excavated circle with the central base removed.

Checking the depth

5 Use a batten to the same depth as a brick and cut to size lengthways.

6 Nail a longer batten to the first as shown. Add a layer of sharp sand to the trench and tamp down whilst going around the circle.

Placing bricks

7 Begin with a circle of half-bricks at the centre. Half-bricks prevent large gaps occurring between each brick, giving a neater and more uniform appearance. Use the string to check each row of bricks is centred.

8 Continue building the circle. Keep checking the correct orientation of the bricks using the string.

Finishing off

9 Complete the circle. Fill any gaps with sand and ballast, and/or a few broken roof tiles.

10 Remove some bricks to make areas for planting. Remove the turf from the centre to avoid a tiny circle of lawn. Either plant it up or infill it with paving or tiles laid on edge.

Left The finished, planted brick circle. Setting the bricks slightly lower than any surrounding lawn will make for easier grass cutting as the mower will glide over the brick edge.

Easy-build shed

This shed replaced one that had fallen down (it had been hand-built just after World War II) and some bits of nice old planking were retained. The replacement shed was less utilitarian, offering a refuge from the elements and a place to brew up tea as well as somewhere to store all the usual tools and gardening paraphernalia. With extra light from well-positioned windows it can now be used for seed sowing and for bringing on plants as well. If you take your time and enlist the help of a buddy, building your own shed is inexpensive, offering a more robust structure than a shop-bought one.

SIZE **2500mm (w) x 2100mm (d) x 2300mm (h)**

Above and **opposite** From tumble-down wreck to a handsome shed in a few days with some hard work and a friend to help. The finished shed is enhanced with a pallet-made window box (see page 122).

Constructing the shed

Make a base and lay decking. It is a quicker option than a slab base and it is easier to fix walls to it. Erect studwork frames, clad with timber and add windows, roof and door.

Materials
Timber for decking and shed (mostly recycled). Studwork, screws, nails, concrete for base, recycled windows, roofing felt. Recycled door and strap hinges. Weatherproof caulking.

Making the base

1 Lay a 'floating' frame for the decking using concrete pads. Here nine pads were laid, providing a supported span of around a metre in each direction. Both new and recycled joists were used. Ground is rarely totally flat, so level the framework using short lengths of offcut timber or bricks placed between pads and joists. Check carefully with a spirit level. All recycled timber should be treated with a wood preservative, paying particular attention to sawn ends. Any short lengths of offcut timber are best immersed in the preservative so that it is fully absorbed into them.

Building on concrete pads

As an alternative to laying a solid concrete base, pads of concrete will usually suffice for relatively lightweight wooden structures. Pads should be placed every metre or so. Dig holes at least 30cm square and at least 30cm deep. Fill the bottom with a layer of hardcore, tamping down firmly and fill each hole with a strong concrete mix of around one part cement to five parts ballast. Allow to dry thoroughly, ideally for at least a day..

2 Lay the decking, which will become the shed floor. The decking here was cheaply bought gravel boards, though recycled decking planks could be used just as easily (see pages 54-57).

Building the shed

3 Make frames for the walls and windows. Studwork timber is cheap, light and, being of a regular size, convenient to use. Position uprights ensuring that the spaces between them allow for the recycled windows to be fitted. Diagonal bracing will keep the frame square during construction.

4 Add timber cladding from whatever boards are available. Here some were used from the original old shed.

5 Position windows as required. Timber-framed windows can be screwed into the studwork walls. Make good any gaps around the windows with fresh putty or weatherproof caulking or filler.

6 To provide a slight slope for the roof, build up one end of the walls with a couple of extra layers of studwork timber. It's not essential but a waterproof breathable membrane can be used on walls, seen here as the grey material.

7 Fix more timbers to provide a support for the roofing panels. It's good to have an overhang to avoid any rain entering above the doors.

8 Continue by adding more cladding and windows. Here an old piece of strand board (osb) quickly covers the rear wall of the shed.

9 Add flat sheets of timber, in this case strand board (osb), for the roof base. Nail or screw the timber sheets to the roofing timbers prior to covering with a layer of roofing felt. This is attached using a heavy duty stapler or short clout nails.

10 Add lengths of timber to cover the ends of the roof rafters. Here an old door was used on its side to fill in the bottom half of one wall.

11 Add roofing felt. The shed is now near completion, just requiring a door, some more cladding and some finishing touches, such as covering the edges of the roofing felt (overleaf).

Finishing touches for the shed

Above A found, two-part, or stable, door was attached using strap hinges. They are a good choice as they are both strong and easily surface mounted.

Left Pallet planks and sections are used to provide simple shelving, and split strips from a blue-painted pallet (far left) make an attractive tool rack with next to no effort.

Above A light and airy space is achieved by including lots of windows. Glass covers large areas quickly and the effect is a great improvement on the small plastic windows in many shop-bought sheds.

Reviving a deck

2

SKILL LEVEL

If the good-quality wood in your deck is beginning to show its age, you can make a low-cost improvement by simply taking up the boards and turning them over. Any that are too badly damaged or rotten should be discarded and replaced with new lengths of similar wood. Always check the sub-frame for any signs of rot, and cut out and replace any rotten lengths with treated timbers. If the joists were originally spaced too far apart, as in this case, you may need to add in some new supports. A simple recycled wood version is shown on page 59.

SIZE **4300mm (w) x 2800mm (l)**

Opposite and **above** The newly laid deck (left) and, a few weeks later, the garden planted up (above). Decking planks have been arranged with the new planks laid in the main section and the original planks laid in the lesser-used 'bridge' and side area (shown clearly in picture left).

Reconstructing the deck

Remove existing decking, check subframe, adding and repairing timbers as required. Fix new and reclaimed decking.

Materials
Decking, timber for rafters, screws or nails, landscaping fabric, wood preservative.

1 Carefully take up existing decking and check subframe. The original decking boards were western red cedar. Any decent lengths were salvaged and pressure washed before re-laying the other side up.

2 Lay out salvaged lengths to ascertain quantities of new timber required. Remember that new decking timber must be exact same thickness as existing decking boards.

3 You may find that some decking joists have rotted or have been placed too far apart. If that is the case, cut out any decayed joists and replace with new timbers. Add joists as necessary, making sure that the space between joists is no more than 40cm. Apply wood preservative.

4 The 'bridge' was remade using a new timber frame wider than the original and raised a little higher to give a distinct change of level. Short bracing timbers are used to give additional strength to the frame and to prevent the deck from flexing.

5 If using decking over soil, put down landscaping fabric to stop weeds from growing underneath. Here butyl liner was also used to establish a bog garden to the side of the bridge.

6 Lay decking boards by screwing or nailing on to joists. If your decking is a lightweight wood such as cedar, then nailing is sufficient. Use a nail punch so that nail heads are just below the level of the decking boards.

7 Trim decking boards – most easily done by using a circular saw against a batten as a guide. Otherwise draw a straight line on the decking in pencil and cut carefully with a hand saw. Sand cut edges with a fine-grade sandpaper.

8 Complete the decking. This shows an awkward little area where a section of sub-frame needed to be added.

Making a simple deck

This deck was laid simply and quickly over an old shed base that was fortunately level and sound. All timber was reclaimed. The decking boards were the grooved variety used other side up.

1 Clean decking boards, by scraping off any residue before using a wire brush and water.

2 Make a subframe using 4 × 2 timbers. Short cross-members will give the frame more rigidity.

3 Small wooden blocks will stop the frame from moving when fixing decking boards.

4 Use a thin sheet of timber between the boards to achieve even spacing between them.

5 The near-completed deck can now be stained or oiled, or simply left to weather down.

Open pallet gazebo

3

SKILL LEVEL

I had acquired a stack of pallets from skips and friendly builders but wasn't sure what to do with them. My initial thought was to make a second shed for my allotment, but it would probably just become a repository for junk, so I decided to make a seating area with some shelter for the occasional barbecue, weather permitting!

SIZE **2400mm (w) x 1400mm (d) x 2100mm (h)**

Left The site for the gazebo is marked out with canes. The area is a bit constricted owing to the nearby apple, pear and plum trees, but they provide a shady canopy with the bonus of a summer harvest at hand.

Opposite The beauty of designing your own structure is that not only is it unique but you can easily add to it if you wish. Here I decided that screening was required to mask out the background clutter. The open pallet ends made the building appear unfinished, so I filled them with twigs, providing a miniature haven for wildlife. The completed gazebo is shown overleaf.

Constructing the gazebo

Start by making a deck (see page 59), construct pallet walls, then fix the walls to the deck, add pergola and any screening the that may be required.

Materials
Pallets, timber for decking, battens, screws, nails, concrete, bricks or paving slabs, angle irons (optional).

Making the deck

1 Constuct a frame for your deck from stout timber joists. This deck was made from old roofing joists and reclaimed decking timber placed grooved side down. (See pages 54-59 for more information on deck construction.) The deck was levelled by placing bricks and paving slabs under the joists, with additional support from angle irons at the deck corners. For added longevity, coat all timbers with a wood preservative.

Making pallet walls

2 For ease of construction choose pallets of the same size. To make a solid pallet wall start by removing some of the timber slats (see also page 10).

3 Insert the removed slats into a second pallet. If you are lucky the slats will fit the spaces neatly. If not, you may have to trim them to fit, which is most easily achieved with an electric jigsaw.

Fixing walls to decking

5 From the back, screw the wall panels together.

6 Using long screws, fix panels to the deck. Drill holes at an angle to ensure screws penetrate both panels and decking.

4 By joining pallet walls together at right angles the structure will have much more strength and stability.

Making shelving

7 In the process of construction you may find that you have acquired quite a few offcuts of pallet. What to do with them? You might decide to make a planter with them (see page 116) or a window box (see page 122). Here I've used them to incorporate some shelved areas on the wall.

8 For the shelved areas place offcut sections of pallets as requred. Screw or nail into position.

9 To give a neater result, finish off open sections of pallets with slats. The shelves are perfect for placing pot plants or lanterns.

Making an overhead screen

10 For your screen you will need a number of battens of a fairly uniform size.

11 An old door makes an ideal frame for the screen. If you don't have one to hand, you can make a simple frame using timber of a suitable size.

12 Fix four battens to your required height using a spirit level to check verticals.

13 Attach a bracket to each of the four uprights. These will hold the frame in position.

14 The roof frame is positioned and screwed to the uprights.

15 The horizontal laths are screwed into the roof frame making sure they are equally spaced.

Making side screens

16 You can now add side screening if, as here, you want to provide more shelter.

17 Strips of pallet wood are used to form the side screen. Place a batten between the strips to ensure uniform spacing and check horizontals with a spirit level.

18 Add a quick-fix screen from salvaged timber stapled to a pallet frame (see opposite, top left).

Finishing touches for the gazebo

Above Additional support for pallet walls, positioned discreetly out of sight.

Above The quick-fix screen made from strips of feather-edge timber salvaged from a derelict shed.

Above A hollow pallet end wall infilled with twigs and sticks.

Right Revitalized decking turned grooved side down and then wire brushed to remove any dirt.

Log path

1

SKILL LEVEL

If you are having, or have had, some trees sawn down, you can use them to make a hard-wearing path, best suited to a wild or country-style garden.

SIZE **800mm wide**

Materials

Logs, sand or ballast, crushed stone or tiles (optional).

Making the path

I If already sawn, choose logs of fairly equal lengths. If sawing specially, then saw each log to an equal length as it will make the path easier to construct. Lay out the logs and number them. Take a picture for reference.

3 Referring back to your picture, reposition the logs in numbered order. Build up soil under any shallower logs to achieve a level top surface.

2 Remove the logs and dig out the soil to the width of the path and to the depth of the deepest log.

4 Backfill around and between the logs with soil.

5 Add a layer of sand or ballast (or both) and firm down using a stout batten.

6 Add a layer of crushed stone or tiles between the logs to give a more decorative appearance, if you wish.

Right The path having weathered after a year or two. Toadstools added to the the effect in autumn. If the path becomes slippery, simply nail some wire netting to the logs.

Retro greenhouse

SKILL LEVEL

I had been considering making a small greenhouse for a while: somewhere to propagate a few trays of seeds and to grow a few tomato plants in pots or growbags. I had squirrelled away some plastic sheets for a fairly functional greenhouse, but when a friend replaced her lovely old leaded windows with double glazing, I seized the opportunity! I had to buy some battens to join these window units together and also some timbers for the deck floor, though I could have used pallet wood (see page 10). The only other costs were for screws, putty, flashing, strap hinges, paint, concrete and ballast; altogether no more than around £50. With a little forethought it could be built in a weekend. When creating the base and footings, remember that the concrete needs to dry overnight.

SIZE **1800mm (w) x 1300mm (d) x 2000mm (h)**

Size The greenhouse that was made measures 180 by 130cm. The size will be determined by the found glass panels. Drawing scale plans on paper will help you decide how to configure the panels (see right). I considered an octagonal greenhouse but the roof construction would have been much more complicated, so I abandoned that idea.

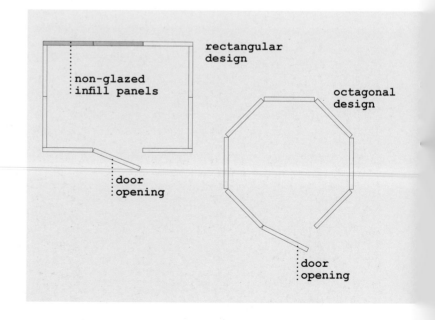

Constructing the greenhouse

This was built by first assembling the glass panels, then creating the foundations for the base and the base itself, then assembling the sides, then the roof and finally adding the door and any finishing touches.

Materials
Found glass, recycled timber, recycled polycarbonate roofing, screws, nails, mastic or putty, primer and paint.

Making up the glass

1 Carefully match and group pairs of windows.

2 Join windows together by screwing wooden battens to each side of window frames. Depending on the window sizes, these battens may have to be longer than the length of the windows for adequate height from floor to ceiling (see diagram).

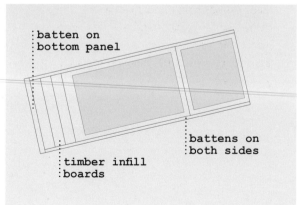

batten on bottom panel

battens on both sides

timber infill boards

Making the base

3 Make a base. An easy and quick solution is to make a wooden deck slightly larger than the finished greenhouse (see page 59). The frame for the base was made from old timbers and levelled with chocks placed on concrete footings (see page 48).

4 Add the boards. This was done with a mix of pallet boards and offcuts of sawn timber. When joining timber don't forget that the joined lengths have to sit on a joist. When all boards have been laid, trim to size.

Building the sides

The greenhouse walls can be of timber or glass, or a mix of the two, as shown.

5 Place window units in position using a spirit level to check verticals. Brace with angled battens to stabilize the structure until it is fixed.

6 Position more units, measuring carefully for the width of the door opening. The greenhouse sides are almost complete. Add cross-timbers and screw together for strength. Clamping the timbers helps to stop any movement whilst drilling and then remove battens. At this stage repair any defective glazing with putty or exterior-quality mastic.

Adding the roof

Cut the roofing material to size, allowing for a slight overhang on all sides. In this case an offcut of polycarbonate sheeting from an old conservatory was used. The sheeting should incline downwards from front to back, to allow rainwater to drain off.

7 Remove the bracing timbers and add the roof. Self-adhesive flashing was used to make the joints between wood and plastic watertight.

Finishing

8 Infill any open areas (here, at the bottom of the rear window) with wooden planks. Paint the exterior with a wood primer followed by a top coat of exterior-quality paint in a colour of your choice. Attach the door with suitably robust hinges (in this case strap hinges). Strap hinges are both heavy duty and easy to attach as they are surface mounted and require no rebating to fix. If the door opening is a little too wide apply a timber batten (below) to the edge of the door and paint to match.

9 Salvaged door-furniture made good, with recycled handles and catches.

10 Paint the interior - ideally white to enhance light levels - using ordinary emulsion paint. Add any shelves as required. The ones here are hinged and have chains at each end to allow them to be folded up, saving space when large plants are grown.

Woodland den

2

This structure was inspired by the frame of an old parasol as its starting point (see below). It had been languishing in my garden for months and I had originally intended to use it as the roof for a little gazebo, but simply turning it upside down led to other ideas. The resulting den is somewhere for children to play and have fun or for adults to escape to. Apart from the parasol frame, a few screws, staples and nails, most of the elements were found from nature.

SIZE 1500mm (w) x 1500mm (d) x 2000mm (h)

Opposite The den could easily be made into a hide by covering the sides in leafy branches. As it is in a wooded araea, the top has been left open although it could easily have been covered too. Don't expect your structure to last for eternity, though it will remain sound for a few years, especially if placed on free-draining land.

Above Lanterns hanging from the den at dusk. Always take great care when using naked flames in the garden and never leave them unattended.

Above The parasol fabric had rotted away, leaving quite a strong wooden framework.

Making the den

Choose a suitable site, make the basic structure and add more elements to it. Complete it with finishing touches to suit.

Materials
Timber poles, string, nails, screws, staples, parasol frame (optional).

Making the basic structure

1 Find a suitable space to build your den. This site uses three closely planted trees with straight stems.

2 Add additional poles. Here another three were added to make a rough hexagonal shape.

3 Dig out 20cm (8in) deep holes, insert poles and backfill with soil, pressing down firmly.

4 Add cross-pieces at a suitable height. Nails will help to hold these pieces in place before tying in.

5 Add more cross-pieces and tie in with stout cord or string. Use a spirit level to check horizontals. Levels don't need to be exact but the structure will look better if not built at odd angles.

Working on the base

6 Make stakes from found materials such a medium-sized branches. Cutting stakes with angled ends will make then easier to hammer into the ground.

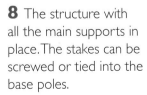

7 Place base pieces at the bottom of each two poles and hammer in the stakes to keep them in position.

8 The structure with all the main supports in place. The stakes can be screwed or tied into the base poles.

Adding uprights

9 To fill in the frame add thinner poles at fairly regular intervals. Screw uprights to horizontals and wrap with string, both for strength and to add to the rustic charm of the structure.

10 Add cross pieces, weaving in and out of the uprights depending on the straightness of the wood. Tie together at some intersections.

11 If you try to bend branches too much they will snap. To avoid this saw them halfway through before bending.

Making an entry

12 Don't forget to leave an opening for access!

13 Stout string is wound around the opening to enhance the impression of an entrance.

14 Using a Y-shaped branch makes an effective addition to the entrance. Small children might appreciate a low entrance height to stop easy access by annoying parents!

15 String or cord is kept taut by kotting an end and using staples to keep it from becoming loose.

Finishing touches

16 Not every side of the den has to be made of twigs or branches. Three sides were made from foliage simply tied into position. As it decays, more fresh foliage can be added. If a plentiful supply of vegetation is not to hand, then camouflage netting is a good alternative, and it's permanent too.

17 The upside-down parasol placed roughly in position. Its supporting spokes create the roof.

18 A child's-eye view. It's always good to look at things from different angles before committing to a final design.

19 The parasol with the upright shortened and the longer parasol spokes tied into position. The 'roof' could be covered with foliage (bracken would be good) or left open, as it is here.

Above The newly completed den. The parasol spokes provide supports for which you can hang lanterns (see page 79).

SCREENS & SUPPORTS

Bamboo screen

Pallet wall

Bamboo arch

Rusty metal arch

Bendy plastic feature

Bamboo screen

You may regard your garden as a tranquil and verdant oasis, but there will be times when you may want to mask an unsightly view or provide some shelter or support for your plants. A screen can be quick to put up and can often be made from materials that are to hand. The one illustrated here was made from stout bamboo poles that had previously provided support for runner beans. As an alternative, an old wooden ladder sawn in half would do the job just as effectively (see page 91).

SIZE 1900mm (w) x 1000mm (h)

Left The bamboo pole screen helps mask the view of a brightly painted shed.

Opposite Viewed from the reverse side, the screen gives glimpses of a mini-orchard beyond, while providing support for grape vines. The primulas in pots on the shelf at its base add a nice touch of colour.

Materials
Stout bamboo poles, string, nails, tape.

Making the screen

1 Begin by nailing into a couple of uprights to support the first pole, insert it and check it is level. Two nails will support each end of the bamboo poles.

2 Use a wooden block to achieve equal spacing. between the poles.

3 Add the other horizontal poles in position. In this example a longer double (top) pole attaches to another structure.

4 To give the screen additional strength, and to prevent it sagging, a couple of diagonal poles are taped into position temporarily.

5 Secure the diagonal poles to the horizontals with thick twine; trim the ends to give a neat finish.

6 To prevent any movement, run a long piece of twine over each pole, twist it around each nail head and tie it at each end.

Above An alternative screen is made from sawn sections of an old wooden ladder which have been screwed together. It has been attached to a low wooden fence and pallet wall to make it sturdier.

Pallet wall

Pallets are a great source of cheap wood and they can be used for a wide range of structures in the garden, more or less as they come in some cases or with the timbers carefully levered apart in others (see page 10). They will always have a very rustic finish to them, but painting them can give a different look. Pick a soft colour that tones well with the garden for the best effect.

Here the the pallets have been stacked double height to create a pallet planter wall, with the pockets between the struts used for small planting or storage areas. They could also be used for decorative displays that double as habitats for insects and other wildlife.

For a wall as shown, choose eight pallets of the same size and thickness, although you can ring the changes to suit your own needs.

SIZE each wall **1500mm (w) x 1500mm (h)**

Right The newly planted wall. Using lengths of guttering (see page 116) allows for plants to be easily arranged and changed as necessary to suit the season. The same wall is shown later in the year (see pages 98-99) planted with pansies, ivies, saxifrage and *Solanum pseudocapsicum*, the latter displaying orange-red berries that last well into winter, unless devoured first by birds!

Constructing the wall

Collect pallets of similar size and ideally the same thickness to give a neat and co-ordinated result. You could make a smaller wall using just four pallets.

Materials
Pallets, nails and screws. Wood stain (to add colour to the wall). Lengths of guttering for planting into (see page 116).

Preliminaries

1 Decide where to position your pallet wall. This one used the panels of an old shed as a back support. If your wall is free-standing, you will have to insert sturdy posts to which to join the pallets for stability.

2 First create a floor base, which could be paving or decking. Here decking was laid using reclaimed decking boards (see page 59). Begin by positioning the first pallets in order to work out the most effective layout.

Colouring timber

3 If you are staining your pallets then it's best to do this before fixing together so that all visible timber is uniformly coloured. It's also much easier to keep your deck or paved floor base unspattered.

4 Brush on the stain; it will normally dry in an a hour or so.

Placing pallets and checking levels

5 Position the first pallet. If your floor isn't properly level, place some slivers of timber or broken tile under one end of the pallet and check with a spirit level. Fix with screws. A diagonal timber from the shed wall to the deck will secure the structure during construction.

6 If pallets have to be cut down to fit, then it gives a neater finish if you cut a middle section rather than an end one.

Cutting pallets to fit

7 Place pallets together to ascertain how much needs cutting off.

8 Then draw a vertical line down the pallet and saw to the required size.

9 Position the cut pallet to check that it fits neatly. Remove it and apply wood stain before fixing finally.

Finishing off

10 For any short sections of pallet wall cut as necessary to give a symmetrical appearance.

11 As well fixing them to a supporting structure, join the pallets to each other using long screws.

12 To avoid too many changes of level along the top, place small blocks of wood on top of the lower pallet in order to raise the upper one.

13 The upper pallet placed in position showing how the wall is now of equal height along its length.

14 To make smaller sections more secure, two short lengths of timber are screwed into the lower pallet. The upper pallet is then inserted and screwed together (after staining all pieces).

Wall planters

15 Short lengths of guttering make ideal planters (see page 116). Remember to paint ends of sections.

16 The guttering planters are easily inserted and stay in place without fixing.

17 For taller plantings, remove a second strut.

Bamboo arch

This is a simple, quickly made structure created from stout bamboo poles and cable ties to support climbers temporarily. The ties allow you to fix the poles together without drilling (which might cause them to split). And, if the ties are left uncut, they act as a decorative feature too. This arch will last for a few seasons only. Use metal rods for a more permanent structure.

SIZE 1200mm max (w) x 2200mm (h)

Below and **opposite** The bamboo arch in the middle of summer, supporting climbing roses and a fast-growing jasmine (see also pages 106-107).

Making the arch

You need a sturdy base to anchor the poles to the ground. Construct tripod frames with stakes and poles joined with horizontal bamboo poles at the top.

Materials
Bamboo poles, stakes, wire and cable ties.

Preparing a base

1 Saw six equal lengths of timber or battens to a point at one end.

2 Use a club hammer to drive the pointed stakes into the ground. Attach a pole to each one using strong wire.

Tying poles together

3 Wire three poles together to make two tripods. Then position the horizontal poles.

4 Adjust horizontal poles as preferred and remove any supports that are no longer necessary.

5 Tie in the horizontal poles to the tripods.

Strengthening the structure

6 Tie in short lengths of thinner bamboo poles across the tripods to strengthen them and to provide a climbing frame for the plants.

7 Add further cross-pieces, both for strengthening and for visual impact. Some trial-and-error will be involved before coming up with an arrangement that looks good.

Finishing off

8 Tie in the climbing plants to the tripods.

9 Replace any temporary wiring with cable ties.

Right The arch frames the entrance to an allotment shed. (See pages 46-53 for construction details.)

Rusty metal arch

SKILL LEVEL 1

On a friend's allotment I spotted a heap of rusty metal (below) in various shapes and sizes. He was surprised that I could even think of any use for it. But a handily bent section of metal tubing led me to devise an arch support for a vigorous blackberry plant. It must have taken no more than half an hour to make, so a good project for a beginner who has few or no tools! It is unlikely to last for many seasons, though.

SIZE 1200mm max (w) x 2000mm (h)

Materials
Sections of rusty metal poles, bits of metal, wire, plant ties.

Right The newly completed arch with some blackberry stems tied in. It's best to tie in such stems when they are long and flexible; shorter stems can easily snap.

Making the arch

1 Decide on the width of the arch. Push the first metal poles into the ground to a depth of around 20cm by hammering the other end with a club hammer.

2 Then hammer in more poles for the other vertical support.

3 Use a heavy-gauge wire to secure the curved top section to the uprights. Twist the wire tightly with heavy-duty pliers.

4 The main structure of the arch is now ready for more upright poles to be added.

5 Another section of metal (the rusted base of a tin drum) is added to the top to define the shape of the arch. Make sure it is well above head height!

6 A rusty metal grill joins sections of poles together.

9 The completed arch. Prune off any long and trailing stems as needed.

7 The structure becomes stronger as each further element is fixed.

Below The arch later the same summer, supporting a heavy crop of ripening fruit.

8 Tie in the plant stems once the structure is complete.

Bendy plastic feature

My new allotment came with an old
rusty metal frame (below) and a shed
roof festooned with bendy plastic
tubing. Too lazy to dig out the metal
frame and take the tubing to the
skip, I turned it into a feature. I
am not sure about its artistic merit,
but it did disguise the frame quite
well! You could use a similar feature
as a climbing plant support, too.

SIZE 1800mm (w) x 2000mm (h)

Materials
Bendy plastic,
wire, short wooden
rods.

Opposite The completed
feature. It started out looking
a little like an eye but later
morphed into a shape a bit
like a teddy bear!
Note The same type of
tubing, cut into very short
lengths, was used to make
the bug hotel (see page
136). The tubing can also be
used to provide the support
for fleece-covered plant
protection.

Making the feature

1 Take a length of plastic tubing, measure it for size and cut to the desired length. As the tubing is usually stored on a large drum, it naturally retains its curved shape.

2 Attach sections of tubing to the chosen frame. A frame could easily be made from lengths of timber too. Fix by drilling small holes in the ends of the tubing and tie in with strong wire.

4 Bring the two lengths of tubing together, and drill through both ends of the tubing and rods.

3 To make a tubing circle, create the shape with the tubing and cut to size where the ends of the tube overlap. Cut a short wooden rod and place in one end of the tube. It needs to be a snug fit.

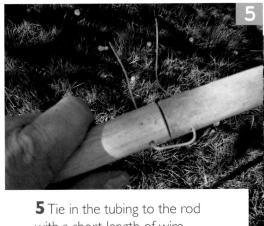

5 Tie in the tubing to the rod with a short length of wire, twisted at its ends. The tube will naturally assume a circular shape.

6 Add the circle and more lengths of tubing to the frame, tying it in place with wire.

7 Shorter lengths of tubing can be used for different shapes. Just drill holes in the tubing and tie together with wire. Trim the ends of the wire.

8 Add more sections. Where you stop is up to you... or when you run out of tubing!

CONTAINERS & RAISED BEDS

Guttering planter

Grass box planter

Window boxes

Hexagonal planter

Planted gabion

Bug hotel

Guttering planter

1

**SKILL
LEVEL**

If you have some offcuts of plastic guttering, rather than skipping it, use it instead as a space-saving planter, ideal for sowing seeds or for growing on small plants.

SIZE 120mm (w) x 60mm (h) x desired length

Materials
Plastic guttering, timber offcuts, screws, rubber from a bicycle inner tube (optional).

Above The planters are used here to grow plants from seed in the greenhouse. Similar planters of a shorter length were used in the construction of the pallet wall (see page 92).

Constructing the planter

1 Using the curve of the guttering as a template, draw an outline on a piece of timber.

2 Using an electric jigsaw, cut out the shape of the guttering profile. You need two for each planter.

3 Cut the timber into two short lengths. The bottom piece will hold the planter in place.

4 Insert the cut timbers into each end of the the gutter. Though not essential, a rubber gasket, made from an old bicycle inner tube, finishes the end off neatly.

5 Trim the rubber gasket. Then drill drainage holes in the bottom of the guttering.

6 Use short screws to fix the wooden ends to the guttering.

7 Attach base timbers by screwing through the guttering into the wood near each end. These will keep the planter stable when planting.

Grass box planter

An old mower grass box was rusting away on a neighbouring plot at my allotment and the owner let me revamp it as a container. I was entranced by the almost obliterated script lettering on the front but I hadn't realized quite how thin and holed the metal had become, so I had to resort to adding a lining of mesh and wire netting. An old oil drum or large food tin cut in half would make just as good a container. Be warned: oil drums when cut with electical tools have been known to explode from the sparks causing vapours inside the container to ignite, so clean them really thoroughly before cutting.

SIZE **400mm (w) x 280mm (d) x 330mm (h)**

Right The container planted up simply for summer with geraniums and trailing lobelia. A plastic liner will stop the compost from drying out too quickly and the sawn branch stand provides a firm support while enhancing the general appearance.

Making the planter

Use wire netting if the box is holed by rust, add a plastic liner and polystyrene chips, and make a support from sawn branches.

Materials
Grass box, plastic sheeting, wire netting, sawn branches, screws, polystyrene chips.

Preparing the grass box

1 Check grass box for any holes.

2 If rust and holes are a problem, line the box with wire netting or mesh.

3 To keep compost in place line the box with plastic. A large plastic bag will be fine for this purpose.

4 A layer of polystyrene chips in the base will stop the container from becoming too heavy when filled with compost.

Making the stand

5 Make two cross-sections from sawn branches screwed together at the centre.

6 Add thinner sawn branches to strengthen.

7 Place the grass box on its support, fill with compost and plant up.

Window boxes

① SKILL LEVEL

Here are two versions of a pallet wood window box: one shallow, one deeper. You can use offcuts of pallet wood to make them. The sections of pallet you have left will determine the form and depth of the window box that you make. For the shallow version, select plants that don't require deep compost and that won't need watering every day.

SIZE 1000mm (w) x 130mm (d) x 150-220mm (h)

Above A simple planting of trailing pelargoniums in a shallow window box, their vibrant red flowers in strong contrast to its deep blue colour.

Making a shallow window box

1 Attach wire mesh to the underside of the pallet section using wide-headed nails or staples.

2 Stain to the required colour using a water-based wood stain.

3 Insert a waterproof liner using butyl or plastic sheeting. Slit the bottom of the liner to provide drainage.

4 Fix a batten to the structure supporting your window box. If attaching it to brickwork, be sure to use suitable fittings.

5 Position the base of the window box on the supporting batten.

6 Screw the window box to the wall. Avoid screwing through the plastic liner as it will catch and tear when the screws are turned.

Making a deeper window box

5 Line the front of window box with hessian or burlap (an old sack will do).

1 A found metal grid fitted the space between the pallet struts. Otherwise use sturdy wire mesh netting cut to size.

2 Mark some cross-pieces of the grid with tape and cut out using a hack saw (to make larger areas for planting).

3 Nail a couple of battens to the underside of the window box to support the weight of the box when filled with compost and plants.

4 Before attaching the metal grid, stain the timber to the chosen colour. Then attach the grid.

6 Add a waterproof membrane to contain the compost; plastic carrier bags are a cheap option. Slit the bags at the base to provide drainage.

Below The deeper box planted at two levels. Planting pockets were made by cutting slits through both burlap and plastic liner for the ivies.

Opposite The shallow window box in close-up.

Above Pallet sections for the shallow window box opposite and the deeper version (above right).

Above (right) The deeper version planted with tobacco plants, cyclamen and variegated ivies, a planting combination that lasts well for much of the summer.

Hexagonal planter

When you have made a few pallet projects you may well end up with a number of offcuts. As I had several leftover pallet ends I considered making a metre-square garden from four offcuts, one for each side, but as I had six sections to spare, the square became a hexagon, and the vegetables became herbs and flowers.

SIZE **2200mm at widest point**

Above and **opposite** The planter consists of six pallet sections, positioned to form a hexagon. The original pallet was already painted bright blue, so I chose the blue, green and mauve planting of herbs and summer-flowering perennials to complement it.

Making the planter

Dig over an area of ground to roughly the chosen dimensions and position the sections of pallet. Level the sections, join together and plant.

Materials
Pallet sections, timber offcut and timber connectors, nails, landscaping fabric.

Placing sections

1 Dig over the site, removing any weeds and place six pallet sections in the shape of a hexagon. Check levels.

2 Adjust angles using a template. This can be cut from a piece of scrap timber. to the angle you require.

Levelling and preparing sections

3 Add small wooden blocks nailed to the underside of the sections if you need to adjust the levels.

4 To suppress weeds, nail a barrier material, such as landscaping fabric, to the underside of each section. This can also be used to line the central section, too.

5 Turn sections back over and place accurately by checking the angle against the template.

Connecting sections

6 Use timber connectors (bought from a DIY store) to join the sections together. Using the template, carefully bend each connector to the correct angle.

7 Nail sections together. When they all connected together, the planter will stay in place.

Planting preparation

8 Fill the central section with soil and/or compost. Level off in central section and plant up. Trim away any landscaping fabric.
Note: This shows the initial planting. More plants were added a few days later and in just a few weeks the planter was in full flower (see page 129).

Planted gabion

I had a number of gabions (industrial metal cages) left over from designing a garden for a flower show. What could I do with them? Some were used in my garden, filled with logs to make low sturdy fences and the remainder I decided to use on my allotment, as an additional space for planting flowers and crops. They also provide a haven for bugs and other wildlife, so I filled some sections with surplus harvest, such as windfall apples, green tomatoes and the woody stems of cardoons.

SIZE each gabion **1000mm (w) x 330mm (d) x 600mm (h)**

Above and **right** The planting, in containers constructed from old drawers, sits on top of the wood-filled gabions. Pansies, chrysanthemums and ivies give a long and colourful display from summer to early autumn.

Filling the gabion

1 Construct the planters from either recycled timber or, more conveniently, from drawers from an old desk, as here.

2 If using drawers, cut them down to size and reassemble, discarding the drawer base.

3 Fill the gabions with waste wood before inserting the planters, ensuring they are level with the gabion top.

Materials

Gabions, scrap timber or recycled drawers, screws or nails, wire, plastic liner.

4 Staple or tack plastic sheet to the sides of the wooden planters. Slit the plastic to provide drainage and fill with soil or compost, ready for planting. Planting is best done after the top section of gabion is filled and wired into place.

Above Sticks can be used to fill in any gaps between gabions.

Opposite Strawberries are a great crop for gabion planting if grown in full sun, and fed and watered regularly. Being raised, the gabions offer good slug protection.

Bug hotel

SKILL LEVEL ①

Bug hotels - an environment to attract beneficial insects to the garden - have become increasingly popular, often made from piles of pallets and filled with all sorts of things. This one is a compact version for the small garden, made from plastic water pipes which I inherited. Don't underestimate how much tubing you will need. This one used up between four and five metres!

SIZE 300mm (w) x 80mm (d) x 320mm (h)

Materials
Timber, plastic tubing, sticks and straw, screws or nails.

Left An old ivy-covered apple tree makes a perfect spot for a bug hotel, which is simply held in place using a length of stout string nailed to the frame. Position the frame in a sheltered spot so it dosen't get blown about in strong winds.

Making the frame

1 Make a frame from equal lengths of timber. Screw or nail pieces together.

2 Cut lengths of tubing. These look better if cut slightly longer than the depth of the frame. They don't have to be all of identical length and can be easily cut with anvil secateurs.

3 Pack cut lengths as tightly as possible. Any larger gaps can be filled with twigs or short lengths of bamboo. Place horizontally on a flat surface and push all tubes down.

4 The completed structure. If not placing it against a flat wall, back the frame with a thin timber sheet.

5 Make your bugs feel more at home by packing some tubes with lengths of straw or dry grass.

Basic tools

The projects were completed using items that you may well have in your tool kit already, especially the essentials such as a hammer, saw, drill, spirit level and measuring tape. Others can be bought, borrowed or hired as needed.

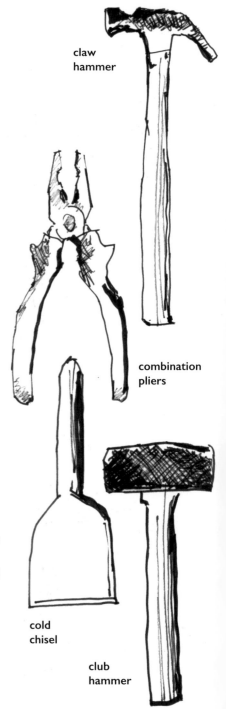

claw hammer

combination pliers

cold chisel

club hammer

Hammer A claw hammer is essential as it can also be used for removing nails. Handles are generally made of wood, steel or fibreglass.

Club hammer A heavy hammer generally used with a **cold chisel** for splitting bricks.

Saw One with fewer teeth per cm is best for sawing rough timber and pallets. Saws with more teeth per cm are used for finer work; cutting will take a lot longer.

Drill A cordless hammer drill is best, especially when working in remote locations. The hammer action is ideal for drilling into brick and masonry. They can also be used with a wide range of **drill bits** and **screwdriver heads**.

Jigsaw An electric jigsaw can be used with a range of blades for cutting different types of material, such as wood, metal, plastic and so on. Coarse-toothed blades can be used for cutting wood along the grain and thinner, fine-toothed blades will cut intricate shapes in timber and other sheet materials.

Circular saw Great for quick cutting of timber, both thick timber joists and sheet materials. When cutting sheet materials, fix a batten to the timber to be cut and run the saw against the batten to ensure a perfectly straight cut.

Crow bar An essential tool for prizing apart pallets (see also page 10). The claw head is useful for removing stubborn nails from timber.

Spirit level Essential for ensuring structures are perfectly horizontal and vertical. Longer ones give more accuracy but shorter versions are best in confined spaces.

Combination pliers Necessary for working with metal, for bending and cutting. For cutting thin sheet metal use a pair of **tin snips**.

Other useful tools A good **craft knife** with replacable blades. A retractable **tape measure** (5m minimum). **Wire brush** for cleaning wood and/or metal. A selection of garden tools including a **rake, fork** and **spade**. **Loppers** and **secateurs** for plant pruning and clearance. **Safety glasses** and **gloves** when working with timber and metal. A **wheelbarrow** for mixing sand and cement.

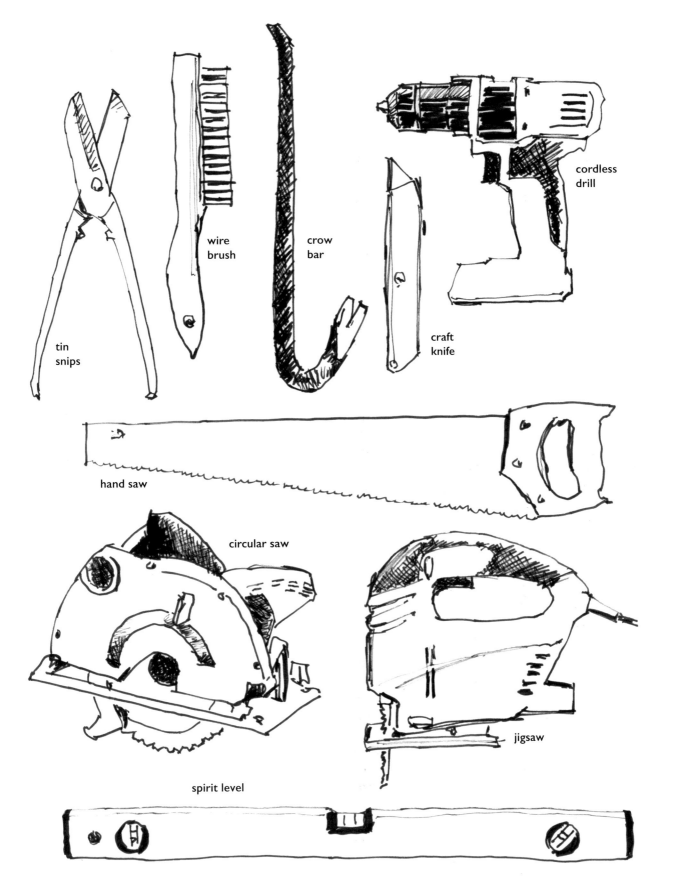

tin
snips

wire
brush

crow
bar

craft
knife

cordless
drill

hand saw

circular saw

jigsaw

spirit level

Adirondack chair templates

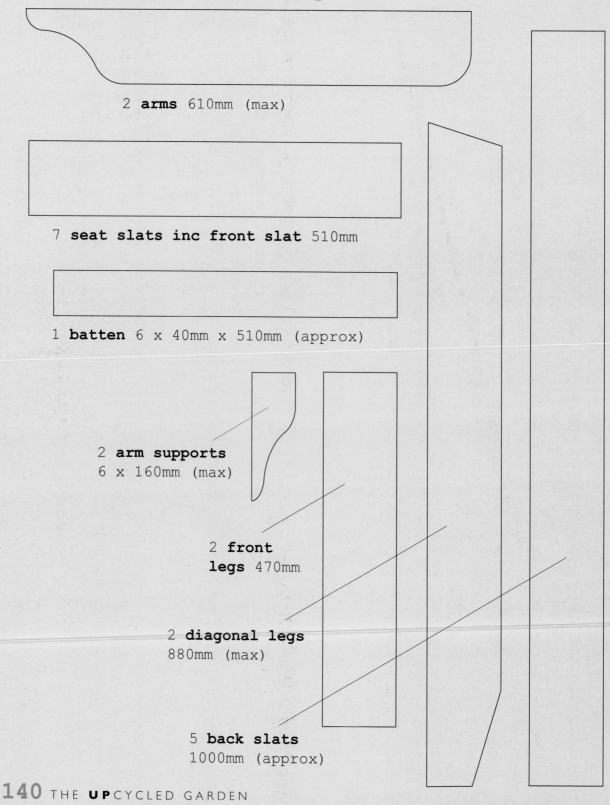

2 **arms** 610mm (max)

7 **seat slats inc front slat** 510mm

1 **batten** 6 x 40mm x 510mm (approx)

2 **arm supports**
6 x 160mm (max)

2 **front**
legs 470mm

2 **diagonal legs**
880mm (max)

5 **back slats**
1000mm (approx)

The Adirondack chair is the only project
in this book where it is best to work
from scale drawings. The illustrations
shown here are at a scale of 1:5. All
timber is a nominal width of 100mm and
a thickness of 20mm approximately with
the exception of a batten, which is
60mm x 40mm (again measurements are
approximate). Once you have made your
first chair you can make some tweaks, such
as altering the length of the chair back
or altering seat or arm heights.

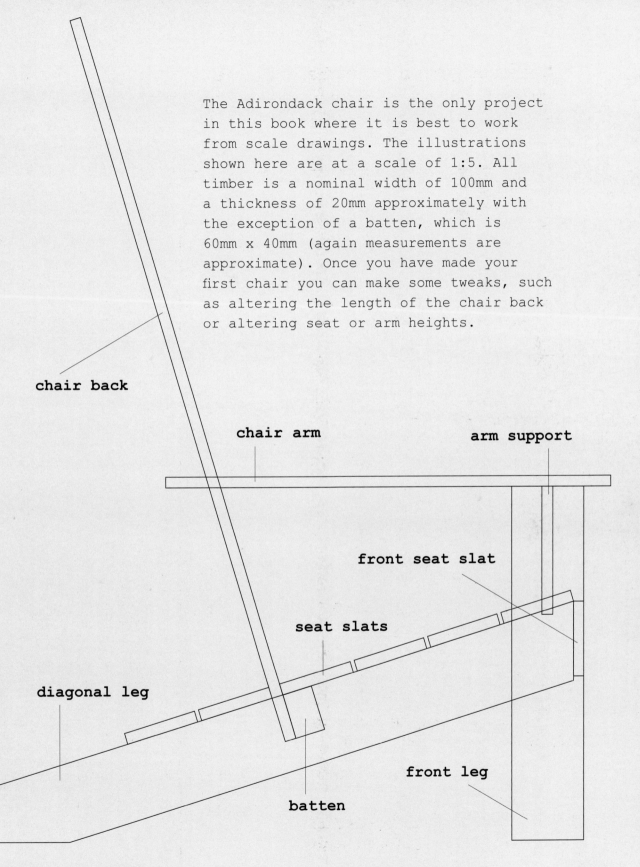

chair back

chair arm

arm support

front seat slat

seat slats

diagonal leg

batten

front leg

Safety precautions

Anyone undertaking DIY projects should pay attention to some basic security precautions. First of all, when using any kind of electrical cutting equipment plugged into a mains source make sure a circuit breaker is employed, to guard against inadvertently cutting through an electricity cable. Again, when using electric cutting equipment or other tools where there is any danger of flying shards or spinters, wear safety googles.

When working at height always ensure your step ladder is firmly positioned. If it feels unsteady, get your builder's mate to hold it in place.

Wear protective (steel toe-capped) boots when lifting or moving heavy materials.

Protect your hands with strong gloves when handling sharp-edged materials.

Make sure there is a first-aid box with basic first aid kit nearby including bandages, plasters, and antiseptic cream and wipes.